SEASON TO SEASON

A Year on the FARM

by Christina Mia Gardeski

PEBBLE
a capstone imprint

Pebble Plus is published by Pebble
1710 Roe Crest Drive, North Mankato, Minnesota 56003
www.capstonepub.com

Library of Congress Cataloging-in-Publication Data is available on the Library of Congress website.
ISBN 978-1-9771-1293-4 (hardcover)
ISBN 978-1-9771-2007-6 (paperback)
ISBN 978-1-9771-1294-1 (eBook PDF)

Summary: From mending fences to seedlings and combines, life on the farm changes from season to season. Discover what farmers do in winter. Learn how crops are harvested in fall. Real-life photographs follow the seasons and capture the beauty of a year on the farm.

Editorial Credits
Elyse White, designer; Jo Miller, media researcher; Tori Abraham, production specialist

Image Credits
Alamy: MediaWorldImages, 19; iStockphoto: Fertnig, 21; Newscom: Danita Delimont Photography/David R. Frazier, 15; Shutterstock: AlinaMD, 9, Brian A Wolf, Cover, (bottom left), Denton Rumsey, 13, Djordje Ognjanovic, 11, JuneJ, 17, MaxyM, Cover, (top right), 3, Michael Shake, Cover, (bottom right), Opapa, 5, periphoto, Cover, (top left), Volodymyr Burdiak, 7

Design Elements
Shutterstock: Alexander Ryabintsev, Minohek

Printed and bound in China.
002493

Note to Parents and Teachers

The Season to Season set supports national science standards related to earth science. This book describes and illustrates how life on a farm changes with the seasons throughout the year. The images support early readers in understanding the text. The repetition of words and phrases helps early readers learn new words. This book also introduces early readers to subject-specific vocabulary words, which are defined in the Glossary section. Early readers may need assistance to read some words and to use the Table of Contents, Glossary, Read More, Internet Sites, Critical Thinking Questions, and Index sections of the book.

All internet sites appearing in back matter were available and accurate when this book was sent to press.

Table of Contents

Spring Is Here!

The ground thaws and turns soft.

A new season begins.

Spring is here! A plow cuts rows

of fresh soil on the fields.

Soon green seedlings sprout.

Farmers clean barns and move livestock outside. Baby animals are born. Piglets drink their mother's milk. At many farms, visitors can pet the animals.

Hello, Summer!

The seasons change.

Hello, summer! Days get longer.

Crops grow in the warm sun.

Tomatoes ripen. Cornstalks

stretch from the earth.

Farmers shear their sheep.
The wool is used to make yarn.
Farmers weed the fields and
check for pests. Insects and
rabbits can harm crops.

Fall Appears!

Soon days grow shorter and cooler. Fall appears! More crops are ready to harvest. A combine collects wheat and other grains.

Some farmers plant winter crops. Spinach and potatoes grow in big greenhouses. Visitors go on hayrides to pick apples and pumpkins.

Welcome, Winter!

It is cold outside. The seasons change again. Welcome, winter! The fields are empty and frozen. The animals stay warm inside farm buildings.

Farmers feed the animals and clean their stalls each day. Farmers fix tools and tractors. Some rebuild fences. Farmers also order seeds.

Soon spring will come again.
The farmer will be ready.
The seasons change four times
each year. And there is always
a job to do on a farm!

Glossary

combine—a large farm machine that is used to gather crops

crop—a plant farmers grow in large amounts, usually for food; farmers grow crops such as corn, soybeans, and peas

greenhouse—a warm building with clear walls in which plants grow

harvest—to collect or pick crops from the field

hayride—a farm ride in an open wagon filled with hay

livestock—animals raised on a farm such as cows, sheep, and pigs

pest—an insect or animal that can harm crops

season—one of the four parts of the year: winter, spring, summer, and fall are seasons

seedling—the young plant that grows out of a seed

shear—to cut off the hair or wool of an animal

thaw—melt

Read More

Hoena, Blake A. *The Farm: A 4D Book.* North Mankato, MN: Pebble Plus, 2018.

Mattern, Joanne. Farm Animals. Washington, D.C.: National Geographic, 2017.

Reader, Jack. *Working on the Farm.* New York: PowerKids Press, 2018.

Internet Sites

John Deere for Kids
https://www.deere.com/en/connect-with-john-deere/john-deere-for-kids/

My American Farm
http://www.myamericanfarm.org/

Critical Thinking Questions

1. How would you describe a farm to someone who has never been to one?

2. Describe some of the foods grown on a farm.

3. Why don't some farm animals live outside in winter?

Index